W0246979

**Sun Tzu** was a Chinese general, military strategist and philosopher who lived during the Eastern Zhou period (771–256 BCE). His seminal work, *The Art of War*, is widely regarded as the most influential book on strategy ever written. Covering principles of preparation, adaptability and understanding of both oneself and one's opponents, Sun Tzu's timeless insights have transcended their military origins to inspire leaders in business, politics and personal development worldwide.

**Gill Hasson** has 20 years' experience teaching and writing on a range of issues to do with personal and professional development, mental health and wellbeing. She is the author of more than 22 books: the bestselling *Mindfulness, Mindfulness Pocketbook, Emotional Intelligence, Positive Thinking,* the *Sunday Times* bestseller *How to Deal with Difficult People*, plus other books on the subjects of resilience, communication skills and assertiveness.

# THE ART OF WAR

## ALSO IN THE SERIES

*Autobiography of a Yogi*
*Meditations*
*The Prophet*
*Think and Grow Rich*

# THE ART OF WAR

## *Timeless Wisdom Distilled*

---

**SUN TZU**

Edited by Gill Hasson

First published in Great Britain by John Murray One in 2025
An imprint of John Murray Press

SRD



The material for *The Art of War* is based on the complete 1908 edition of *The Book Of War: The Military Classic of the Far East*, translated by Captain E. F. Calthrop, R. F. A., published by John Murray and now in the public domain. This edition is not sponsored or endorsed by, or otherwise affiliated with Everard Calthrop, his family or heirs.

A CIP catalogue record for this title is available from the British Library

Hardback ISBN 978 1 399 82150 6
ebook ISBN 978 1 399 82151 3

Typeset by KnowledgeWorks Global Ltd.

Printed and bound in India by Manipal Technologies Limited, Manipal

John Murray Press policy is to use papers that are natural, renewable and recyclable products and made from wood grown in sustainable forests. The logging and manufacturing processes are expected to conform to the environmental regulations of the country of origin.

John Murray Press
Carmelite House
50 Victoria Embankment
London EC4Y 0DZ

John Murray Press
123 S. Broad St., Ste 2750
Philadelphia, PA 19109

www.johnmurraypress.co.uk

John Murray Press, part of Hodder & Stoughton Limited
An Hachette UK company

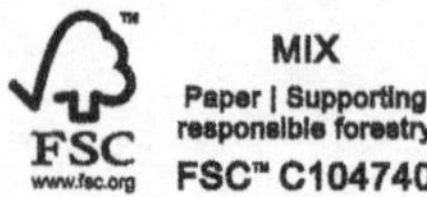

# Contents

# Preface

*The Art of War* by Sun Tzu was written over two thousand years ago during a particularly disordered time in Chinese history known as the Warring States period (476–221 BCE) when, in what is now eastern China, more than 20 rival states fought for dominance.

Each state was ruled by its own sovereign, each intent on reunifying China under their rule, as emperor. The success of each ruler was dependent on a warlord – a commanding general. One such commanding general was Sun Tzu, whose military success informed his treatise *The Art of War.* Over time, other experts – military leaders and writers on military affairs – added their commentaries and explanations to Sun Tzu's directives.

The title of *The Art of War* emphasizes the creative approach and skill required for a military leader in terms of preparation, strategy and tactics. Although the treatise was written so long ago, the principles and methodology of *The Art of War* have, over the centuries, remained relevant because of its focus on strategy and tactics.

## War as last resort

The premise of *The Art of War* is that a war is best won without any fighting at all. It is only if all other means have failed that a full-on confrontation should be turned to; war must be a last resort.

War, says Sun Tzu, is an expensive business and so a long-drawn-out campaign should be avoided. He asserts that, when war is prolonged, no one benefits. We can see how, today, this advice applies to legal battles, too. A protracted divorce, for example, is costly both financially and emotionally. In this example, mediation is often recommended as a more positive alternative.

However, if war cannot be avoided, it should be fought, on both strategical and psychological levels, to keep damage and destruction to a minimum.

Sun Tzu's approach is a combination of peaceful actions and aggressive warfare, an approach that reflects the Taoist principles of yin and yang – forces that are both opposing and complementary.

As well as stressing the importance of avoiding prolonged conflict, Sun Tzu stresses the need for preparation. He explains how meticulous preparation allows a leader to strike fast when an opportunity presents itself.

Further advice covers a wide range of issues, including the use of deception: for example, to appear strong when you are weak and appear weak when you are strong. There is also an emphasis on the need to consider environmental conditions: to be able to react quickly to changes in the weather, for example, and to adapt tactics to specific situations.

Sun Tzu explains that a group is only as strong as its leader and emphasizes the importance of acting with integrity, leading by example, being kind yet strict, and treating captured troops and defeated soldiers with respect.

## Relevance of *The Art of War* today

*The Art of War* is one of the most influential leadership books of all time. Its advice on how to use strategy and skill, cunning, tactics and discipline has become a source of inspiration across many spheres such as business, law, espionage, politics, competitive sports, and video and board games.

In fact, what *The Art of War* teaches can be applied in just about any situation in which there is the potential for conflict or competition and a need to outwit an opponent.

## About this book

*The Art of War* is traditionally divided into 13 chapters, each one devoted to a different set of skills related to leadership, military strategy and tactics. Each chapter is comprised of numbered sentences.

The first annotated English translation, under the title *The Book of War*, was completed and published by Everard Ferguson Calthrop (1876–1914) – a lieutenant-colonel in the British Army and distinguished linguist who served in the Far East – in 1908.

In this book – *The Art of War: Timeless Wisdom Distilled* – each chapter starts with a brief summary of its contents. Then, on each left-hand page, you will read a short passage from Calthrop's original translation, together with his useful annotations and, on occasion, other notes. A small number of passages have been omitted where they add very little. On each facing page there is an explanation of Calthrop's translation of Sun Tzu's original text.

Together, all 13 chapters come together to teach today's reader how to be a successful leader, how to avoid an outright confrontation, and how to outwit the opposition.

## CHAPTER ONE

# PRELIMINARY RECKONING

SUN TZU BEGINS his treatise by stating that there are five principles to be taken into account when planning and preparing for war, along with seven elements to be considered.

He points out that a leader should also be aware of any further circumstances that could be of benefit and provide an advantage; that whatever their plans (which "must not be made known beforehand"), a leader must remain flexible in order to respond advantageously to changes in circumstance or conditions that occur.

In the final part of the chapter, Sun Tzu explains his assertion that "War is a thing of pretence" (often stated as "All warfare is deception") and concludes with advice on the need to change tactics according to the enemy's condition.

" Now, in war, besides stratagem and the situation, there are five indispensable matters. The first is called The Way;* the second, Heaven; the third, Earth; the fourth, the Leader; the fifth, Law.

The *Way* or the proper conduct of man. If the ruling authority be upright, the people are united: fearless of danger, their lives are at the service of their Lord.

*Heaven*. Yin and Yang;† heat and cold; time and season.

*Earth*. Distance; nature; extent; strategic position.

The *Leader*. Intelligence; truth; benevolence; courage and strictness.

*Law*. Partition and ordering of troops.

These things must be known by the leader: to know them is to conquer; to know them not is to be defeated. "

---

* The five virtues of humanity, righteousness, propriety, wisdom and faith are known as The Way.

† Yin and Yang are the two principles into which natural phenomena are divided in Chinese philosophy. Yin is the masculine, active or light principle, and Yang is the feminine, passive or dark principle. In this connection, day and night, rain, mist and wind are designated.

SUN TZU OUTLINES five principles that govern warfare and must be fully understood by any leader seeking victory.

1. The Way refers to unity and shared purpose between the leader and the people. When the people are in agreement with the leader's vision, they are more likely to support him, even in the face of danger.

2. Heaven is concerned with environmental factors: the weather, seasons and time, all of which can significantly influence the outcome of military engagements. Knowledge of seasonal variations can influence the planning of campaigns, while weather conditions can affect troop movements and morale.

3. Earth involves the physical terrain. A thorough knowledge of the landscape allows leaders to manoeuvre effectively, choose advantageous positions and anticipate the enemy's movements.

4. The personal qualities of the leader are essential for effective leadership, inspiring confidence among troops and making wise strategic decisions.

5. The fifth principle – Law – refers to the organization of troops: the structuring of ranks and enforcement of discipline. A well-disciplined and orderly force is more likely to carry out complex manoeuvres and adapt to changing circumstances.

These five principles serve as a strategic framework for assessing both one's own position and that of the enemy. "To know them is to conquer; to know them not is to be defeated."

"Further, with regard to these and the following seven matters, the condition of the enemy must be compared with our own. The seven matters are:—The virtue of the prince; the ability of the general; natural advantages; the discipline of the armies; the strength of the soldiers; training of the soldiers; justice both in reward and punishment. Knowing these things, I can foretell the victor."

Sun Tzu outlines a further seven key factors that a leader must consider in both their own and the enemy's forces in order to predict the outcome of a conflict. These considerations serve as a strategic framework for evaluating relative strengths and weaknesses:

1. The virtue of the prince concerns the moral integrity and leadership qualities of the sovereign. A ruler who embodies righteousness and commands respect fosters unity and commitment among his people.
2. The ability of the general refers to the competence, wisdom and strategic acumen of the leader.
3. Natural advantages refer to the geography and terrain. Understanding and making good use of natural features can provide a tactical edge in warfare.
4. The discipline of the armies can determine the reliability and effectiveness of the forces.
5. The strength of the soldiers concerns the physical abilities, morale and resilience of the troops.
6. Training of the soldiers requires regular and rigorous training to ensure that soldiers are prepared for combat.
7. Justice in reward and punishment refers to a fair and consistent system of discipline, incentives and rewards in order that soldiers are motivated and perform their duties diligently.

By using these seven principles to compare one's own forces and those of the adversary a leader can predict the likely victor in a confrontation.

> " If a general under me fight according to my plans, he always conquers, and I continue to employ him; if he differ from my plans, he will be defeated and dismissed from my service. Wherefore, with regard to the foregoing, considering that with us lies the advantage, and the generals agreeing, we create a situation which promises victory; but as the moment and method cannot be fixed beforehand, the plan must be modified according to circumstances. "

In this passage, Sun Tzu asserts that a leader who adheres to the planned strategy is likely to achieve victory and retain command. Conversely, a leader who deviates from the established plan will be both defeated and dismissed.

However, Sun Tzu also acknowledges that the specifics of warfare cannot always be predicted and so plans must be adapted according to changing circumstances. The challenge for a leader is therefore to ensure that their actions remain aligned with the key aims while also responding effectively to real-time developments on the battlefield.

This balance is essential for achieving sustained success in military operations.

" War is a thing of pretence: therefore, when capable of action, we pretend disability; when near to the enemy, we pretend to be far; when far away, we pretend to be near. Allure the enemy by giving him a small advantage. Confuse and capture him. If there be defects, give an appearance of perfection, and awe the enemy. Pretend to be strong, and so cause the enemy to avoid you. Make him angry, and confuse his plans. Pretend to be inferior, and cause him to despise you. If he have superabundance of strength, tire him out; if united, make divisions in his camp. Attack weak points, and appear in unexpected places. These are the secrets of the successful strategist, therefore they must not be made known beforehand.

At the reckoning in the Sanctuary before fighting, victory is to the side that excels *in the foregoing matters*. They that have many of these will conquer; they that have few will not conquer; hopeless, indeed, are they that have none. "

SUN TZU'S ASSERTION that "war is a thing of pretence" highlights the fact that success often hinges on deception: misdirection and psychological manipulation. He lists a number of deceptive tactics, the aim of which is to unbalance the opponent and force them into reactive and disadvantageous positions.

Sun Tzu insists that the tactics must be both kept secret and planned for in advance. Although he doesn't explain how to carry out the deceptions he describes, it would seem reasonable to assume that the use of spies, so clearly explained in Chapter 13, plays a crucial role in misleading and deceiving an adversary.

> Sun Tzu's concept of war as an exercise in deception remains a key element across a number of domains. Most recently, for example, we have all become familiar with the term "fake news". Fake news refers to false or misleading information presented as legitimate news. It is often created deliberately as a method of deception, to influence public opinion or damage reputations. Fake news can take various forms, including fabricated stories, hoaxes, propaganda or manipulated media, and is frequently disseminated via social media platforms, websites or even traditional news outlets.

## CHAPTER TWO

# OPERATIONS OF WAR

SUN TZU BEGINS this chapter by pointing out that war is an expensive business. His key advice is to avoid a long-drawn-out campaign because when war is prolonged, no one benefits. He suggests that "He who does not know the evils of war will not reap advantage thereby", and concludes that the aim should be victory, not lengthy campaigns.

“Now the requirements of war are such that we need a thousand light chariots with four horses each; a thousand leather-covered chariots, and one hundred thousand armoured men; and we must send supplies to distant fields. Wherefore the cost at home and in the field, the entertainment of guests, glue and lacquer for repairs, and necessities for the upkeep of waggons and armour are such that in one day a thousand pieces of gold are spent. With that amount a force of one hundred thousand men can be raised:—you have the instruments of victory.”

SUN TZU EXPLAINS the substantial economic cost of warfare, detailing the extensive resources – transport, troops and provisions – required to sustain a large military campaign. He emphasizes the daily expenses associated with such an operation – the entertainment of guests, repairs and maintenance of equipment. However, Sun Tzu recognizes that although expensive, if it is possible to fund all that is needed to go to war, "you have the instruments of victory".

" But, even if victorious, let the operations long continue, and the soldiers' ardour decreases, the weapons become worn, and, if a siege be undertaken, strength disappears.

Again, if the war last long, the country's means do not suffice. Then, when the soldiers are worn out, weapons blunted, strength gone and funds spent, neighbouring princes arise and attack that weakened country. At such a time the wisest man cannot mend the matter.

For, while quick accomplishment has been known to give the victory to the unskilful, the skilful general has never gained advantage from lengthy operations.

In fact, there never has been a country which has benefited from a prolonged war. "

SUN TZU WARNS of the detrimental effects of extended military engagements. Prolonged conflict depletes resources and exhausts the troops. Then, in your weakened position you are open to attack from enemy states. Even the most wise and clever of leaders cannot then turn things round.

Sun Tzu advocates for swift, decisive actions in warfare. He acknowledges that some of the least skilled leaders have been known to achieve success. However, wise and skilled leaders have never gained from a long-drawn-out campaign and no country has ever benefitted.

> Prolonged conflict rarely leads to successful outcomes. This principle remains relevant today, as history has shown that drawn-out wars often result in significant economic, political and human costs without guaranteeing success. The same is true for legal disputes – rarely does either side benefit from protracted disagreements.

> "He who does not know the evils of war will not reap advantage thereby. He who is skilful in war does not make a second levy, does not load his supply waggons thrice."

SUN TZU ASSERTS that only a leader who is fully aware of the evils of war – the inherent challenges and costs – can understand the best way to proceed. Once war is declared, a wise leader doesn't wait for reinforcements or fresh supplies. Rather, they move forward, without delay.

This highlights the importance of thorough preparation and efficient resource management. A skilled leader ensures that right from the start, their army is well supplied, minimizing the need for additional levies or repeated resupply efforts, which can drain resources and morale.

" War material and arms we obtain from home, but food sufficient for the army's needs can be taken from the enemy. The cost of supplying the army in distant fields is the chief drain on the resources of a state: if the war be distant, the citizens are impoverished.

[...]

Therefore the intelligent general strives to feed on the enemy; one bale of the enemy's rice counts as twenty from our own waggons; one bundle of the enemy's forage is better than twenty of our own. "

SUN TZU'S ADVICE to obtain food supplies from the enemy emphasizes a strategic approach to minimize the logistical and economic burdens of warfare.

Transporting provisions from one's own country not only strains resources but also diminishes the army's operational efficiency. In contrast, securing supplies locally from the opposition helps lessen the financial and logistical pressures on the leader's own country.

Why waste time, effort and resources securing food supplies from home when it takes far less effort to steal supplies from the other side?

This advice further supports the need for swift, decisive action in warfare. Sun Tzu advocates for strategies that minimize duration and resource expenditure, with foraging from the enemy being a key tactic in achieving this efficiency.

"Incitement must be given to
vanquish the enemy.
They who take advantage of the
enemy should be rewarded.
They who are the first to lay their
hands on more than ten of
the enemy's chariots should
be rewarded;
[...]
The accompanying warriors must
be treated well, so that, while the
enemy is beaten, our side increases
in strength."

IN THIS PASSAGE, Sun Tzu reiterates the significance of morale, efficient use of resources and psychological tactics in warfare.

Rewarding soldiers who secure early gains – capturing enemy chariots for example – encourages further bravery and initiative among troops. Adding captured arms and equipment to your own forces increases your military strength and morale and further weakens the opposition's strength and morale.

Sun Tzu suggests that treating captured soldiers well can lead to their cooperation, further reinforcing one's own ranks.

In modern times, by applying the principles of rewards and the possession of competitors' resources, individuals and businesses can increase their own strength and success. In business, for example, rewarding employees for their efforts and achievements can both motivate them and improve productivity.

Additionally, a business can increase its success by using a competitor's resources to its advantage by, for example, hiring their employees or buying their assets.

" Now the object of war is victory;
not lengthy operations, even
skilfully conducted.
The good general is the lord of the
people's lives, the guardian of the
country's welfare. "

SUN TZU'S ASSERTION that in war the aim is victory, not a long-drawn-out campaign, once again emphasizes the need for swift and decisive action in warfare.

Sun Tzu highlights the immense responsibility borne by military leaders, their decisions having a direct impact on the lives of citizens and the fate of a nation.

> These principles remain relevant today, stressing as they do the importance of efficient planning and the ethical conduct of leadership in a range of contexts such as business and education.

## CHAPTER THREE

# THE ATTACK BY STRATAGEM

IN THIS CHAPTER, Sun Tzu advocates for a strategy that minimizes damage and destruction, and suggests that "the supreme art is to subdue the enemy without fighting". Subduing the enemy's soldiers without fighting also carries the benefit of keeping one's own forces intact.

However, where confrontation becomes unavoidable, Sun Tzu gives advice on what to do if one side is not equally matched in numbers with the opposing side.

Sun Tzu points out the three ways in which a ruler (that is, the government or sovereign of the state) can bring misfortune upon an army, and continues by outlining five conditions in which victory is most likely to be the result.

" Now by the laws of war, better than defeating a country by fire and the sword, is to take it without strife.
Better to capture the enemy's army intact than to overcome it after fierce resistance.
Better to capture the 'Lu,' the 'Tsu' or the 'Wu' whole,* than to destroy them in battle.
To fight and conquer one hundred times is not the perfection of attainment, for the supreme art is to subdue the enemy without fighting.
Wherefore the most skilful warrior outwits the enemy by superior stratagem; the next in merit prevents the enemy from uniting his forces; next to him is he who engages the enemy's army; while to besiege his citadel is the worst expedient.
A siege should not be undertaken if it can possibly be avoided. For, before a siege can be commenced, three months are required for the construction of stages, battering-rams and siege engines; then a further three months are required in front of the citadel, in order to make the 'Chuyin.'†
Wherefore the general is angered, his patience exhausted, his men surge like ants against the ramparts *before the time is ripe*, and one-third of them are killed to no purpose. Such are the misfortunes that sieges entail. "

---

* The Chinese army consisted of 12,500, the "lu" of 500, the "tsu" of 50 and the "wu" of 5 men.

† The "Chuyin" was a large tower or work constructed to give command over the interior of the enemy's fortress. High poles were also erected, from the top of which archers, each encased in an arrow-proof box and raised by a rope and pulley, shot at the besieged.

IN THIS PASSAGE, Sun Tzu outlines a hierarchy of military strategies to maximize the chances of achieving victory with minimal conflict and resource expenditure. The skill is firstly, to outwit the enemy with a superior strategy. The next step is to isolate them from each other and their allies. The third most favorable approach is to confront the opponent's forces in battle. The least favorable policy is to lay siege to a city.

Sun Tzu points out that sieges require extensive preparation, which can deplete resources and morale. The leader's patience wears out, which results in a premature attack and consequently a significant number of casualties. Success is not guaranteed.

> Sun Tzu's advice remains relevant in a number of circumstances concerning conflict in both one's personal and professional lives. His insights encourage us to seek efficient and effective solutions that minimize harm and maximize success.

“Therefore the master of war causes the enemy’s forces to yield, but without fighting; he captures his fortress, but without besieging it; and without lengthy fighting takes the enemy’s kingdom. Without tarnishing his weapons he gains the complete advantage. This is the assault by stratagem.”

THIS PASSAGE SUMS up the need to prioritize intelligence, diplomacy and psychological tactics over direct confrontation. By disrupting the enemy's plans, deceiving them, sowing discord and exploiting weaknesses, Sun Tzu suggests that a leader can achieve their aims without the costs associated with prolonged warfare.

> In contemporary contexts, this approach extends beyond the battlefield. In business, politics, law and personal interactions, success so often relies on foresight, strategic planning and adaptability rather than sheer force.

" By the rules of war, if ten times as strong as the enemy, surround him; with five times his strength, attack; with double his numbers, divide. If equal in strength, exert to the utmost, and fight; if inferior in numbers, manœuvre *and await the opportunity*; if altogether inferior, offer no chance of battle. A determined stand by inferior numbers does but lead to their capture. The warrior is the country's support. If his aid be entire, the country is of necessity strong; if it be at all deficient, then is the country weak. "

IN THIS PASSAGE, Sun Tzu outlines specific actions depending on the comparative strength of forces. His advice serves as a guide for leaders to make calculated decisions based on their strengths compared to those of the enemy.

Sun Tzu then goes on to remind us of the critical role of the military in a nation's fortunes: he explains that the military are the defender, the protector and the strength of the state. If their abilities are strong then the country will be strong. But if their abilities are lacking, the state will be weak.

> Sun Tzu's principles offer valuable advice to us in our personal and professional lives. Assessing one's strengths and weaknesses relative to the challenges of a situation and making informed, strategic choices are key to achieving success while minimizing risks.

“Now a prince may embarrass his army in three ways, namely:—

Ignorant that the army in the field should not advance, to order it to go forward; or, ignorant that the army should not retreat, order it to retire.

This is to tie the army as with a string.

Ignorant of military affairs, to rule the armies in the same way as the state.

This is to perplex the soldiers.

Ignorant of the situation of the army, to settle its dispositions.

This is to fill the soldiers with distrust.

If the army be perplexed and distrustful, then dangers from neighbouring princes arise. The army is confounded, and offered up to the enemy.”

Sun Tzu cautions that a ruler can inadvertently undermine his army through three critical mistakes: firstly, by not understanding when the army should advance or retreat. Secondly, by attempting to govern the army in the same way he governs his people. Thirdly, by being ignorant of the circumstances of the army, failing to place them in the right situation at the right time.

As a result of these errors, says Sun Tzu, the army is confused and distrustful and falls into disarray. It then becomes vulnerable to attack from rulers of neighboring states.

The importance of informed, adaptable and context-sensitive leadership remains relevant to our lives today. Effective command requires an understanding of both the circumstances and the unique nature of, for example, a community, a sports team, a business or organization.

“There are five occasions when victory can be foretold:—
When the general knows the time to fight and when not to fight; or understands when to employ large or small numbers; when government and people are of one mind; when the state is prepared, and chooses the enemy’s unguarded moment for attack; when the general possesses ability, and is not interfered with by his prince.
These five things are the heralds of victory.”

Sun Tzu identifies five conditions under which victory can be anticipated:

1. when the leader knows the time to fight and when not to fight
2. when he knows when to employ large or small numbers
3. when the leader has the support of both the government and the people
4. when the state is prepared and waits to take the enemy unprepared
5. when the leader has military control and has no interference from the ruler or head of state.

These principles stress the importance of strategic planning, unity and skilled leadership in achieving victory.

“It has been said aforetime that
he who knows both sides has
nothing to fear in a hundred fights;
he who is ignorant of the enemy,
and fixes his eyes only on his own
side, conquers, and the next time
is defeated; he who not only
is ignorant of the enemy,
but also of his own resources,
is invariably defeated.”

SUN TZU'S RENOWNED principle asserts the critical importance of understanding both yourself and your enemy. A leader who understands both their own abilities, strengths and weaknesses and those of their opponent can anticipate challenges, make informed decisions and plan accordingly.

Conversely, a leader who lacks knowledge of either side can make miscalculations and assumptions that will lead to defeat.

> The principle of knowing both yourself and your enemy extends beyond the battlefield, offering guidance in various aspects of life, including business, sport, politics and one's personal life. By developing self-awareness and seeking to know and understand others, individuals, teams and organizations can also anticipate challenges, make informed decisions and plan accordingly.

## CHAPTER FOUR

# THE ORDER OF BATTLE

IN THIS CHAPTER, Sun Tzu asserts the importance of protecting yourself and putting yourself beyond defeat; that when you do attack, it should be sudden and swift.

Sun Tzu says that the "clever fighter" excels in winning with ease – that is, by making no mistakes. The "victorious strategist" only declares war once all the necessary plans and preparations ensuring a successful victory have been made.

" The ancient masters of war first made their armies invincible, then waited until the adversary could with certainty be defeated. The causes of defeat come from within; victory is born in the enemy's camp. Skilful soldiers make defeat impossible, and further render the enemy incapable of victory. But, as it is written, the conditions necessary for victory may be present, but they cannot always be obtained. "

In this passage, Sun Tzu's insight teaches that success relies on strategic patience and ensuring one's own position is unassailable – not open to attack – before exploiting the weaknesses of others.

He asserts that security against defeat lies in your own hands. The opportunity to defeat the enemy is provided by the enemy themselves.

> In contemporary contexts – in business, for example – this approach remains relevant. A business is well advised to focus on strengthening its core competencies and market positions, waiting for competitors to falter – be it through mismanagement or market missteps – before making strategic moves to capture market share.

" If victory be unattainable,
we stand on the defensive;
if victory be sure, we attack.
Deficiency compels defence;
super-abundance permits attack.
The skilful in defence crouch,
hidden in the deepest shades;
the skilful in attack push to the
topmost heaven.*
If these precepts are observed,
victory is certain. "

* Literally 9th heaven, and 9th earth. The Chinese divided the earth and sky each into 9 strata.

Sun Tzu's assertion – "If victory be unattainable, we stand on the defensive; if victory be sure, we attack" – emphasizes the strategic balance between offence and defence based on one's current abilities and circumstances.

If victory is uncertain, adopting a defensive stance and staying hidden preserves resources and avoids unnecessary risks. In contrast, if victory is certain, one is well advised to make an immediate attack. A powerful offence that is both swift and overwhelming leaves little time for the opponent to respond.

Once again, the essence of Sun Tzu's teaching is clear: carefully assess both your own and your opponent's capabilities and the external environment to determine whether to adopt a defensive position or seize the initiative offensively.

"A victory, even if popularly proclaimed as such by the common folk, may not be a true success. To win in fight, and for the kingdom to say, 'Well done,' does not mark the summit of attainment. To lift an autumn fleece* is no proof of strength; the eyes that only see the sun and moon are not the eagle's; to hear the thunder is no great thing."

* An animal's coat is thinnest in autumn.

SUN TZU'S ASSERTION that the ability to predict an obvious victory is nothing to be proud of emphasizes the fact that superficial acclaim does not equate to genuine strategic achievement. He warns against mistaking public praise for legitimate skill in warfare.

He further illustrates this with metaphors: "To lift an autumn fleece is no proof of strength; the eyes that only see the sun and moon are not the eagle's; to hear the thunder is no great thing", reiterating his assertion that stating the obvious and achieving easy wins does not demonstrate exceptional skill or insight.

“As has been said aforetime, the able warrior gains the victory without desperate and bloody engagements, and wins thereby no reputation for wisdom or brave deeds. To fight is to win, for he attacks only when the enemy has sown the seeds of defeat. Moreover, the skilful soldier in a secure position does not let pass the moment when the enemy should be attacked.”

IN THIS PASSAGE, Sun Tzu suggests that a skilled leader is one who not only wins but excels in winning with such ease that no one thinks that highly of them or credits them for being clever or courageous. True mastery in warfare lies not in seeking glory through battle but in achieving objectives with minimal conflict.

Once again Sun Tzu's insights remind us that victory is not solely about aggressive action but about strategic patience and ensuring one's own position is unassailable before exploiting the weaknesses of others.

" The army that conquers makes certain of victory, and then seeks battle.
The army destined to defeat, fights, trusting that chance may bring success to its arms.

[...]

The army that conquers as against the army destined to defeat, is as a beam against a feather in the scales. The attack of conquering forces is as the outburst of long-pent-up waters into sunken valleys. "

Sun Tzu asserts the need for thorough planning and preparation before engaging in conflict rather than trusting to chance.

The vivid analogy – "The army that conquers [...] is as a beam against a feather in the scales. The attack of conquering forces is as the outburst of long-pent-up waters into sunken valleys" – emphasizes the power of a well-prepared army.

## CHAPTER FIVE

# THE SPIRIT OF THE TROOPS

In this chapter, Sun Tzu explains that managing a large force is no different from managing a smaller force – it simply requires dividing everyone up into organized units, each unit with its own leader issuing commands and giving direction. Having matched the right person to the right role, one is then able to know the power of a force as a whole rather than the strength of individuals.

The use of direct and indirect manoeuvres is advised. As well as the potential combinations of indirect force being limitless, Sun Tzu says that direct and indirect approaches "give rise to an endless series of manoeuvres. The direct and the indirect lead on to each other in turn [...] who can exhaust the possibilities of their combination?"

In Chapter 1 Sun Tzu famously asserts that "War is a thing of pretence". In this chapter he gives an example, explaining that "simulated disorder postulates perfect discipline; simulated fear postulates courage; simulated weakness postulates strength".

And so the power of a well-organized force, timely decisions and swift action is exponential – or, as he puts it in one of his frequent poetic similes, "as the momentum of a round stone rolled down a mountain thousands of feet in height".

"The control of large numbers is possible, and like unto that of small numbers, if we subdivide them.
By means of drum, bell and flag, the direction of large forces in battle is possible, and like unto the direction of small forces."

In this short passage, Sun Tzu explains how, with proper organization and communication, the challenges of managing large groups can be effectively addressed. By dividing a large army into smaller manageable units, each with its own leader, the commander has effective control of the whole force.

In ancient times, drums, bells and flags served as essential communication methods on the battlefield, allowing leaders to convey orders and coordinate movements effectively. The drum was used to beat the assembly and in the advance, the bell as a signal to halt. Flags were of two kinds, signaling flags and distinguishing banners.

In modern-day warfare, armed forces use communication technologies such as radio, satellite systems and digital networks to coordinate large-scale operations, reflecting the same principles of subdivision and signaling.

“By the skilful interchange of normal and abnormal manoeuvres are the armies certainly preserved from defeat.

[…]

Moreover, in battle the enemy is engaged with the normal and defeated by the abnormal force.*

The abnormal force, skilfully handled, is like the heaven and earth, eternal; as the tides and the flow of rivers, unceasing; like the sun and moon, for ever interchanging; coming and passing, as the seasons.

There are five notes; but by combinations, innumerable harmonies are produced. There are but five colours;† but if we mix them, the shades are infinite. There are five tastes, but if we mix them there are more flavours than the palate can distinguish.‡

In war there are but two forces, the normal and the abnormal; but they are capable of infinite variation. Their mutual interchange is like a wheel, having neither beginning [n]or end. They are a mystery that none can penetrate.”

* The normal and the abnormal refer to what in modern phrase are termed the frontal or holding force and the flanking or surprise force.

† The traditional five Chinese colours are: red, blue, yellow, black and white.

‡ The five cardinal tastes are[:] acridity, bitterness, sourness, sweetness and saline.

Sun Tzu counsels for the strategic importance of combining "normal" conventional, direct tactics with "abnormal" indirect tactics to secure victory.

Normal tactics are the conventional, direct movements that engage the enemy head-on. Abnormal tactics are the indirect approaches such as ambushes, diversions or sudden flanking movements that can surprise and outwit the enemy.

Sun Tzu explains that "the enemy is engaged with the normal and defeated by the abnormal force", highlighting the effectiveness of deception and surprise in warfare. He likens the skillful use of abnormal forces to the continual changes of natural phenomena such as the tides and the flow of rivers, cyclical movements of the sun, the moon and the seasons. Just as these natural elements are constant yet ever-changing, so too should military tactics evolve to meet the demands of each unique situation.

He elaborates on the dynamic interplay between normal and abnormal forces by drawing parallels between the basic elements of music, color and taste, explaining that although there may be just two forces – the normal and the abnormal – they are capable of infinite variation. Their interplay is likened to a wheel, symbolizing continuous motion and the seamless transition between strategies.

"As the rush of rock-shouldering torrents, so is the spirit of the troops. Like the well-judged flight of the falcon, in a flash crushing its quarry, so should the stroke be timed. Wherefore the spirit of the good fighter is terrifying, his occasions sudden; like the stretched cross-bow, whose string is released at the touch of the trigger."

Sun Tzu's vivid metaphors – comparing the spirit of troops to the exponential power of torrential waters, the precision of a falcon's strike and the sudden release of a drawn crossbow – emphasize the necessity of overwhelming momentum, precise timing and the value of being prepared and poised to act instantly when the situation demands.

These principles extend beyond the battlefield. Sun Tzu's insights remind us that in a variety of situations, success often hinges on preparation, timing and the effective use of energy.

" In the maze and tumult of the
battle, there is no confusion;
in the thick of action the battle
array is impenetrable.
If discipline be perfect, disorder
can be simulated; if truly bold,
we can feign fear; if really strong,
we can feign weakness.
We simulate disorder by subdivision;
fear, by spirit; weakness,
by battle formation.
We set the enemy in motion by
adopting different formations to
which he must conform. "

IN THIS PASSAGE, Sun Tzu explains the strategic value of deception in warfare. He tells us that pretending disorder requires perfect discipline; that feigning fear necessitates true courage; and that appearing to be weak demands real strength.

The skill is in deceiving the enemy – to mislead them so that they mistakenly respond to what they *think* you are doing, so creating opportunities that are to your benefit.

The principles of deliberate deception are applicable in a number of contexts where lulling your opponent into a false sense of security can provide a strategic advantage. For example, in a situation where two parties are in negotiation, one side might feign weakness to gain leverage, prompting the opposing side to make concessions.

" If we offer the enemy a point of advantage, he will certainly take it: we give him an advantage, set him in motion and then fall upon him. Wherefore the good fighter seeks victory from spirit, and does not depend entirely upon the skill of his men. He is careful in his choice, and leaves the rest to battle force; yet, when an opening or advantage shows, he pushes it to its limits. As a log or rock which, motionless on flat ground, yet moves with ever-increasing force when set on an incline, so await the opportunity, and so act when the opportunity arrives. If the general be skilful, the spirit of his troops is as the impetus of a round stone rolled from the top of a high mountain. "

SUN TZU COUNSELS for the strategic use of deception to influence the enemy's behavior, telling us that if you appear to give the enemy an advantage, they will take it and then you can strike.

He continues by likening the onset of troops to a log or rock rolling down a mountain. The effect is exponential – a process that builds on itself, becoming more and more powerful. Once again the emphasis is on patience, preparation and the ability to act decisively when the right moment arises.

## CHAPTER SIX

# EMPTINESS AND STRENGTH

IN THIS CHAPTER, Sun Tzu points out that the first army to arrive at the battlefield has the advantage; they will be fresh for the fight while their opponent will arrive exhausted.

He describes a variety of ways to undermine the strength of an enemy before a confrontation takes place, with – as usual – an emphasis on confusing and deceiving the enemy. This will lead them to divert and divide their forces in order to cover each position, thereby weakening them.

Secrecy – remaining unseen and unheard – is crucial. Sun Tzu's advice is to know as much about an opponent's intentions and abilities without revealing anything in return. Knowing the enemy's strengths and weaknesses is particularly important if you are not yet fully prepared for battle.

Sun Tzu likens military tactics to the nature of water: "The shape of water is indeterminate; likewise the spirit of war is not fixed." His advice is not to repeat the tactics which have resulted in a past victory but to modify tactics according to your opponent's intentions and abilities and thereby succeed in winning.

"To be the first in the field, and there to await the enemy, is to husband strength.
To be late, and hurrying to advance to meet the foe, is exhausting.
The good fighter contrives to make the enemy approach; he does not allow himself to be beguiled by the enemy.
By offering an apparent advantage, he induces the enemy to take up a position that will cause his defeat; he plants obstructions to dissuade him from acting in such a way as to threaten his own dispositions.
If the enemy be at rest in comfortable quarters, harass him; if he be living in plenty, cut off his supplies; if sitting composedly awaiting attack, cause him to move.
This may be done by appearing where the enemy is not, and assaulting unexpected points.
If we go where the enemy is not, we may go a thousand leagues without exhaustion.
If we attack those positions which the enemy has not defended, we invariably take them: but on the defence we must be strong, even where we are not likely to be attacked.
Against those skilful in attack, the enemy does not know where to defend: against those skilful in defence, the enemy does not know where to attack."

THIS PASSAGE DESCRIBES key principles of strategic advantage through initiative, positioning and deception. Sun Tzu asserts that whoever has prepared well and is first in the field sets the pace and dictates the way forward. Whoever is second will have arrived in haste, be stressed, drained of energy and be reactive rather than proactive.

The other side can be lured to approach or obstructed in ways that will hold them back.

Sun Tzu suggests that if your opponent's forces are resting, provoke and harass them. If they appear to have plenty, cut off their supplies.

Asserting the importance of appearing at points the enemy must hasten to defend and arriving in places where you are not expected highlights the strategic advantage of innovation and surprise.

Sun Tzu reminds us to attack where they are weak but defend those places of your own that are unlikely to be attacked.

He concludes by emphasizing that a leader who is skilled in attack is one whose opponent does not know what to defend, and a leader who is skilled in defence is one whose opponent does not know where to attack.

“Now the secrets of the art of offence are not to be easily apprehended, as a certain shape or noise can be understood, of the senses; but when these secrets are once learnt, the enemy is mastered.

[…]

Again, when we are anxious to fight, but the enemy is serenely secure behind high walls and deep moats; we attack some such other place that he must certainly come out to relieve. When we do not want to fight, we occupy an unfortified line; and prevent the enemy from attacking by keeping him in suspense. By making feints, and causing the enemy to be uncertain as to our movements, we unite, whilst he must divide.

We become one body; the enemy being separated into ten parts. We attack the divided ten with the united one. We are many, the enemy is few, and in superiority of numbers there is economy of strength.”

In this passage, Sun Tzu asserts the fact that effective offensive strategies are not always obvious. He gives two examples of tactics that exemplify the principle of an indirect approach.

Firstly, rather than engaging in a direct assault on a fortified position, Sun Tzu advocates for feints: attacking a different, more vulnerable target. This compels the enemy to abandon their stronghold to defend the new threat, thereby creating an opportunity to engage them under more favorable conditions.

Secondly, he advocates for deterrence through ambiguity. By presenting a seemingly vulnerable position, one can instill uncertainty in the enemy, causing hesitation and preventing them from initiating an attack. This strategy makes use of psychological warfare, using the fear of the unknown to deter enemy action.

Sun Tzu explains that these tactics – feigned attacks and misleading manoeuvres – force the enemy to spread their forces thin. This division weakens their overall defence, allowing for concentrated and decisive action against isolated units.

" The place selected for attack must be kept secret. If the enemy know not where he will be attacked, he must prepare in every quarter, and so be everywhere weak.

[...]

Everywhere to make preparations, is to be everywhere weak. The enemy is weakened by his extended preparations, and we gain in strength. "

IN THIS SHORT passage, Sun Tzu's advice is to keep intentions hidden so that the enemy is obliged to divide their forces and be stationed in several different places, and as a result, be weak in all places. He explains that a dispersed enemy is vulnerable, allowing a concentrated force to strike decisively at a chosen point and overwhelm them.

The uncertainty of where an attack may occur can erode morale and induce strategic errors, as the enemy attempts to guard against all possibilities.

" Having decided on the place and day of attack, though the enemy be a hundred leagues away, we can defeat him.
If the ground and occasion be not known, the front cannot help the rear; the left cannot support the right, nor the right the left, nor the rear the front.

[...]

If the enemy be many in number, prevent him from taking advantage of his superiority, and ascertain his plan of operations. Provoke the enemy and discover the state of his troops; feint and discover the strength of his position. Flap the wings, and unmask his sufficiency or insufficiency. By constant feints and excursions, we may produce on the enemy an impression of intangibility, which neither spies nor art can dispel. "

Once again, Sun Tzu emphasizes the value of meticulous planning. By determining the best time and place for engagement, a leader ensures that their forces are well prepared to engage even against distant foes. However, if time and place are not known, coordination fails and units cannot support one another effectively.

Even if the enemy is stronger in numbers, it's possible to prevent them from taking advantage of their situation if their plans and chances of success are uncovered in advance.

Sun Tzu's advice is to provoke the enemy to break cover in order to gauge the extent of their activity or inactivity and identify their strengths and weaknesses.

He goes on to suggest that one "[f]lap the wings, and unmask his sufficiency or insufficiency", meaning that by simulating movements or threats on a number of fronts, one can assess the enemy's readiness and resources.

Sun Tzu concludes this passage by suggesting that this constant use of deceptive tactics will confuse the enemy and cause them to think you are powerful and untouchable.

" The general makes his plans in accordance with the dispositions of the enemy, and puts his hosts in motion; but the multitude cannot appreciate the general's intention; they see the signs of victory, but they cannot discover the means. "

SUN TZU ASSERTS the nuanced and often inscrutable nature of effective leadership and strategy. He tells us that a skilled leader makes plans and adapts tactics according to the enemy's tendencies and abilities.

Although the troops see the signs of victory they can't necessarily understand the steps that will lead to success, thus highlighting the importance of trust in leadership.

> Sun Tzu's teachings remind us that effective leadership often involves making informed decisions, the logic of which may not always be immediately apparent to others but which are crucial for long-term success.

“ If a victory be gained by a certain stratagem, do not repeat it. Vary the stratagem according to circumstances.

An army may be likened to water.

Water leaves dry the high places, and seeks the hollows. An army turns from strength and attacks emptiness.

The flow of water is regulated by the shape of the ground; victory is gained by acting in accordance with the state of the enemy.

The shape of water is indeterminate; likewise the spirit of war is not fixed.

The leader who changes his tactics in accordance with his adversary, and thereby controls the issue, may be called the God of war.

Among the five elements* there is no settled precedence; the four seasons come and go; the days are long and short; and the moon waxes and wanes. So in war *there is no fixity*. ”

---

* Wood, fire, earth, metal and water.

IN THIS PASSAGE, Sun Tzu's analogy of water emphasizes the importance of strategic fluidity.

In a confrontation, his advice is to be as water running from high ground to low – that is, taking the path of least resistance. And, just as flowing water shapes the lie of the land, a leader should plan success according to the disposition of their opponent.

Just as water retains no shape, there are no constant conditions and circumstances. If you can modify your tactics according to your opponent, you will achieve your aims.

Comparing the nature of running water to planning military strategy principle emphasizes that there is no fixed method to achieve victory. Sun Tzu's advice is not to repeat the tactics that have gained a victory in the past. Instead, success lies in the ability to assess each unique situation and adapt accordingly.

> This approach extends beyond warfare. For example, businesses that remain flexible and adapt their business plans in response to market changes and consumer behavior are more likely to succeed. Similarly, in our personal lives, individuals who remain flexible and can adapt to change are more likely to do well in a variety of situations in life.

## CHAPTER SEVEN

# BATTLE TACTICS

THIS CHAPTER BEGINS with the somewhat daunting declaration that there is "nothing more difficult" than battle tactics. Somehow, you have to deceive the enemy into thinking that you are a long way off, when in fact you are nearby.

There is a dilemma: to outmanoeuvre the enemy necessitates a fully equipped army, but such a heavily armored army will move slowly. Sun Tzu's solution is to "[m]ove only if there is a real advantage to be gained". He advises that "to concentrate or to divide your troops, must be decided by circumstances" and further advises that in order to traverse unknown terrain, local guides must be hired.

Sun Tzu reiterates the need for speed and secrecy. He suggests that "he who uses his soldiers well, avoids the time when the [enemy's] spirits are keen; but attacks the enemy when he is languid or seeking his camp".

Sun Tzu emphasizes the value of strategic timing and the conservation of one's strength, advocating for engagements that exploit the enemy's weaknesses rather than confronting their strengths head-on.

The chapter finishes with advice on tactics to follow when your opponent is on the retreat.

"The general receives orders from his lord; assembles and settles harmony among the forces, and takes the field.
There is nothing more difficult than Battle Tactics. Their difficulty lies in the calculation of time and distance, and the reversal of misfortune.
To make the enemy take a circuitous route by a show of gain, and then, whilst starting after him, to arrive before him, is to be a master of the art of manoeuvre."

SUN TZU BEGINS by advising that once orders from the state's rulers are received, harmony and respect between the forces is established and the deployment of troops can take place.

He warns us that nothing is more difficult than tactical manoeuvring: a leader must accurately assess distances and timing to outpace the enemy. They must also have the ability to turn misfortune into gain.

Once again, Sun Tzu advocates for the use of deception to influence the enemy's movements, advising that the skill lies in presenting a false advantage – making the enemy take an indirect route – starting after them but reaching the destination before them.

" The operations of an army may reap advantage; the wrangles of a multitude are fraught with peril.

Employing our whole force at one time in order to gain advantage over the enemy, we may not have time enough to gain our object. If we push on with a portion of the force only, the transport is lost. Discarding helmet and armour; stopping neither day nor night; marching double distance; doing double work; and finally contending with the enemy at a distance of a hundred leagues: results in the loss of the general. Since the strong men arrive first, and the tired drop in rear, only one-tenth of the forces is available.

A forced march of fifty leagues to secure an advantage may result in failure to the leader of the vanguard, for only half his men will arrive. After a forced march of thirty leagues to secure an advantage, only two-thirds of the army will be available.

Further, a lack of ammunition, of supplies, or of stores, may lead to disaster. "

In this passage, Sun Tzu highlights the necessity of balancing speed with sustainability and warns of the perils of forced marches. He emphasizes that while speed can offer tactical advantages, it must be balanced against the physical limitations of troops and the necessity of maintaining equipment.

Sun Tzu provides specific examples to illustrate the dangers:

1. Marching 100 li (around 30 miles) means that the strong men arrive first but the others, exhausted, fall to the back, and so the force is scattered and only a small number of your men are available to fight.

2. Marching 50 li (about 15 miles) will result in only half the troops arriving on time.

3. Marching 30 li (around 9 miles) allows for two-thirds of the army to arrive, indicating a more sustainable pace.

These scenarios highlight the fact that excessive haste can deplete an army's strength, leading to potential defeat.

> The principles outlined by Sun Tzu extend beyond a military context. In business and sport, pushing individuals and teams to deliver rapid results without sufficient resources, support or rest can result in stress, low morale and burnout.
>
> Patience and a focus on sustainability are far more likely to deliver success and maintain the wellbeing and productivity of individuals and teams.

"The ruler who is ignorant of the
designs of neighbouring princes,
cannot treat with them.
He who is ignorant of mountain
and forest, defile and marsh,
cannot lead an army.
He who does not employ a guide,
cannot gain advantage from
the ground."

SUN TZU ASSERTS that a leader must possess a clear understanding of both the political landscape and the physical terrain. He explains that the ruler of a state shouldn't enter into alliances unless they know their neighbors' intentions. Without this knowledge, a ruler cannot engage in diplomacy or anticipate threats and opportunities.

He further advises that a leader shouldn't lead their army on a march until they know the lay of the land. Ignorance of the terrain can lead to logistical challenges, misdirected campaigns or ambushes. Employing local guides who are familiar with the area provides a strategic advantage.

“ Disguise your movements; await a favourable opportunity; divide or unite according to circumstance. Let your attack be swift as the wind; your march calm like the forest;* your occupation devastating as fire. In defence, as a mountain rest firm; like darkness impenetrable to the enemy. Let your movements be swift as the lightning. Let as many as possible take part in the plunder: distribute the profit from the captured territory. ”

* This passage was written on the standard of Takeda Shingen [1521–73], one of Japan's most famous generals.

Sun Tzu emphasizes the need to disguise or hide your intentions, to move only if there's an advantage to be gained and for the decision to concentrate or to divide your troops to be decided by circumstances.

His vivid metaphors – to be swift as the wind, calm like the forest, devastating as fire and immovable like a mountain – illustrate the importance of strategic mastery through adaptability, deception and decisive action.

Sun Tzu's advice to distribute the spoils of war among soldiers emphasizes the importance of equitable rewards in maintaining troop morale and cohesion.

“So he who understands the
crooked and the straight
way conquers.
These are the methods of
Battle Tactics.
[...]”

SUN TZU'S ASSERTION, "So he who understands the crooked and the straight way conquers", reiterates one of his key principles: that success in warfare is dependent on the effective use of both direct and indirect approaches.

The direct approach involves conventional, frontal attacks. Although it's straightforward, it is anticipated by the enemy. The indirect approach involves tactics such as deception, diversions and ambushes, all of which mislead and outwit the enemy, striking where they least expect.

In modern times, the use of direct and indirect methods is pertinent in a number of situations, such as business, sport and politics. Understanding when to apply direct action and when to use more subtle, indirect tactics can be crucial in negotiations, competitive strategies and leadership.

"In the morning the spirits are keen; at midday there is a laziness; in the evening a desire to return. Wherefore, he who uses his soldiers well, avoids the time when the spirits are keen; but attacks the enemy when he is languid or seeking his camp. Thus should the nature of energy be turned to account.

To oppose confusion with order, clamour with quiet, is to have the heart under control. To await an enemy from a distance, to oppose hunger with satiety, rest with fatigue, is the way to husband strength. Do not attack where lines of banners wave, nor the serried ranks of battle spread, but patiently await your time."

IN THIS PASSAGE, Sun Tzu emphasizes the value of strategic timing and the conservation of one's strength, advocating for engagements that exploit the enemy's weaknesses rather than confronting their strengths head-on.

He advises leaders to avoid engaging the enemy in the morning when their morale and energy are at their peak, and instead to strike when they are fatigued or less alert, such as at midday or in the evening.

Sun Tzu tells us that by maintaining a calm and disciplined approach a leader can best respond to the chaos of battle. He emphasizes the strategic advantage of patience and preparedness – by allowing the enemy to exhaust themselves through long marches or inadequate supplies, they can be engaged when their morale and physical strength are diminished.

He further cautions against a direct confrontation with a well-ordered, well-prepared enemy and instead to wait for an opportunity when the enemy is less organized or vulnerable in some way.

“Do not attack an enemy on high ground, nor one who has high ground at his back. Do not pursue an enemy who is imitating flight; do not attack a spirited enemy. If the enemy offer an allurement, do not take it. Do not interfere with an enemy who has struck camp, and is about to retire. When surrounding an enemy, allow him an outlet. Do not press a desperate enemy.”

IN THIS PASSAGE, Sun Tzu counsels against direct confrontations when conditions are unfavorable and advocates for a clear understanding of the enemy's mindset and circumstances.

He advises against advancing uphill against the enemy, or opposing the enemy when they come downhill, as to have their back against elevated terrain offers them tactical advantages.

He further cautions against pursuing an enemy that appears to be fleeing or to be offering you an advantage of some sort as it could be a trap. He also advises against attacking a force who are at the top of their game.

Sun Tzu concludes with a warning to recognize the dangers of desperation and advises against preventing the enemy from returning home. An enemy with no escape route may fight with unexpected ferocity. Allowing an outlet can prevent them from becoming dangerously desperate.

## CHAPTER EIGHT

# THE NINE CHANGES

THIS CHAPTER BEGINS with advice on what to do and not do when faced with five different situations and continues with five more specific circumstances which should be avoided. These circumstances relate to the "Nine Grounds" which are further discussed in Chapter 11.

However, Sun Tzu points out that knowing the "Nine Grounds" is not enough; the need to vary tactics according to circumstances is – as stated in previous chapters – crucial.

He asserts that high expectations must be managed and unexpected advantages exploited. Provoke your opponent, keep them busy, bait them and deceive them. Be fully prepared and totally secure.

In conclusion, Sun Tzu describes the "five faults in a leader [which] are disastrous in war" and suggests that "[t]he overthrow of the army and the slaughter of the general arise from them".

" Do not camp on marshy or low-lying ground; enter into friendly relations with neighbouring states; do not linger in a far country; use stratagem in mountainous and wooded country; on death ground, fight. There are always roads that must be avoided; forces that must not be attacked; castles that must not be besieged; ground that must not be chosen for encounter; orders from the lord that must not be obeyed. "

SUN TZU WARNS against camping on unfavorable terrain – marshy or low-lying ground – because of the risk of disease and logistical challenges. He suggests that in mountainous and wooded areas, where direct confrontation may be problematic, strategies such as deception and indirect tactics can provide an advantage. When troops are in a situation where there is no escape – "death ground" – they must be prepared to fight with utmost ferocity, as retreat is not an option.

Sun Tzu's advice to enter into friendly relations with neighboring states allows for secure supply lines. He cautions against lingering in distant lands as it can deplete resources and morale.

Sun Tzu concludes this passage with a list of situations to avoid:

1. Roads that must not be followed: they may lead to ambushes or logistical nightmares.

2. Forces that must not be attacked: that is, enemy forces that are well prepared or in an advantageous position.

3. Castles that must not be besieged: fortifications that would require disproportionate effort to capture.

4. Ground that must not be chosen for encounter: terrain that offers no strategic benefit or poses significant risk.

5. Orders from the lord that must not be obeyed: commands that, if followed, would compromise the mission.

" The general who knows the Nine Changes understands the use of troops; on the contrary, he who does not understand them, can make no use of his topographical knowledge.

In the management of armies, if the art of the Nine Changes be understood, a knowledge of the Five Advantages is of no avail.

The wise man considers well both advantage and disadvantage.

He sees a way out of adversity, and on the day of victory to danger is not blind. "

WHILE SUN TZU does not explicitly list the "Nine Changes", they refer to a set of strategic variations that a leader must understand in order to effectively manage troops and respond to changing battlefield conditions. Knowledge of these variations informs effective utilization of troops. Conversely, without an understanding of the nine changes, knowledge of the terrain may prove insufficient. He further advises that if the nine changes are understood, a knowledge of the five advantages is of no added value.

Sun Tzu concludes this passage by pointing out that a wise leader considers both advantages and disadvantages in every situation. They see a way out of difficulties and even on the day of victory are aware that problems could still arise.

" In reducing an enemy to submission, inflict all possible damage upon him; make him undertake useless adventures; also make neighbouring rulers move as you would desire them by tempting them with gain. Wherefore in the conduct of war do not depend on the enemy's not coming, but rely on your own preparations; do not count on the enemy not attacking your fortress, but leave nothing undefended. "

SUN TZU ADVOCATES for direct action – to "inflict all possible damage" – and indirect action – to make the enemy undertake "useless adventures" – in order to reduce them to submission. He also suggests offering benefits to neighboring allies so as to influence their support.

Sun Tzu then advises against relying on the enemy not coming, but on your own readiness to receive them. Don't rely on them not attacking but rather on the fact that you have made your position unassailable.

In essence, he counsels a combination of vigilant preparation and strategic manipulation to achieve success in warfare.

" Generals must be on their guard against these five dangerous faults:—

[1] Blind impetuosity, which leads to death.

[2] Over-cautiousness, which leads to capture.

[3] Quick temper, which brings insult.

[4] A too rigid propriety, which invites disgrace.

[5] Over-regard for the troops, which causes inconvenience.

These five faults in the leader are disastrous in war. The overthrow of the army and the slaughter of the general arise from them. Therefore they must be carefully considered. "

Sun Tzu identifies five personality traits that can undermine a leader's effectiveness and lead to military disaster.

1. "Blind impetuosity" – engaging in battles without first carefully assessing the risks can lead to unnecessary losses.
2. Excessive caution or fear – this can be paralyzing, resulting in missed opportunities and potential defeat.
3. Having a quick temper – this means that an easily provoked leader risks making rash decisions that the troops view with contempt.
4. "A too rigid propriety, which invites disgrace" – this refers to a leader conforming too strictly to their personal values and so making decisions based on personal honor rather than strategic necessity, leading to a loss of respect.
5. "Over-regard for the troops" – while the troops' welfare is important, excessive concern is a distraction and can hinder decisive action.

When an army is overthrown and its leader killed, the cause can be found among these five faults. The leader must reflect on whether they are prone to any of them.

> Understanding and mitigating these traits is crucial for effective leadership, not only in warfare but also in modern contexts such as business and politics.

## CHAPTER NINE

# MOVEMENT OF TROOPS

SUN TZU BEGINS by offering tips on encamping an army and then continues with advice on the movement of soldiers facing an approaching enemy in four different types of terrain.

This is followed by a description of observable signs that indicate the enemy might be employing deceptive tactics. A column of dust, for example, indicates chariots are on the move, whereas dust that is low and dispersed alerts you to infantry movement. Observations of signs in nature are also important – the rising of birds in flight, for example, is a sign of an ambush.

As well as keeping a close watch on the enemy for signs of their confidence and strength, Sun Tzu lists the signs that show the enemy is flagging.

The chapter ends with advice on the use of discipline and punishment of soldiers.

" Touching the disposal of troops and observation of the enemy in relation to mountain warfare:—
Cross mountains and camp in valleys, selecting positions of safety.
Place the army on high ground, and avoid an enemy in high places.
In relation to water:—
After crossing waters, pass on immediately to a distance. When the enemy is crossing a stream, do not meet and engage him in the waters, but strike when half his force has passed over. Do not advance on an enemy near water, but place the army on high ground, and in safety.
Do not fight when the enemy is between the army and the source of the river.
With regard to marshes:—
Cross salty marshes quickly; do not linger near them.
If by chance compelled to fight in the neighbourhood of a marsh, seek a place where there is water and grass, and trees in plenty in the rear.
In open country place the army in a convenient place with rising ground in the right rear; so that while in front lies death, behind there is safety.
Such is war in flat country.
[...]. "

IN THIS PASSAGE, we learn that terrain is a silent but powerful player in warfare; understanding and adapting to the landscape can decide battles before they begin.

Sun Tzu gives advice with regard to mountainous terrain, flat land, marshes and rivers. He urges leaders to always seek high ground, avoid natural obstacles when possible, use enemy vulnerabilities (like crossing rivers) and position troops with both offence and retreat in mind.

He counsels for battle to take place from a high vantage, to seek stability amidst insecure surroundings, and to identify safe places of retreat before committing to action.

Sun Tzu is also speaking beyond logistics. He is making essential points about psychological preparedness and strategic thinking: position yourself in a way that turns geography into advantage, understand your environment, respect its logic and adjust your plans accordingly.

To understand the terrain is to understand how conflict may unfold.

“ As a rule, the soldiers prefer high ground to low. They prefer sunny places to those the sun does not reach.

If the health of the troops be considered, and they are encamped on high and sunny ground, diseases will be avoided, and victory made certain.

If there be rising ground, encamp on its sunny side and in front of it; for thereby the soldiers are benefited, and the ground used to our advantage.

If, owing to rains in the upper reaches, the river become turbulent, do not cross until the waters have quieted.

Steep and impassable valleys; well-like places; confined places; tangled impenetrable ground; swamps and bogs; narrow passages with pitfalls:—quickly pass from these, and approach them not. Cause the enemy to approach near to them, but keep yourself from these places; face them, so that the enemy has them in his rear. ”

SUN TZU PROVIDES valuable advice on further decisions directed by observing your environment. He advises to look for high, well-lit spaces that allow strategic clarity, influence good morale and support good physical and mental wellbeing.

Good decision making involves an awareness of risk and patience; to wait until a raging river calms before crossing, to avoid energy-sapping terrains like swamps and bogs – and to cause your enemy to stumble into those same traps.

"If there be near to the army, precipices, ponds, meres, reeds and rushes, or thick forests and trees, search them thoroughly. These are places where the enemy is likely to be in ambush.
When the enemy is close, but quiet, he is strong in reliance on natural defences.
If the enemy challenge to fight from afar, he wishes you to advance.
If the enemy be encamped in open country, it is with some special object in view.
Movement among the trees shows that the enemy is advancing. Broken branches and trodden grass, as of the passing of a large host, must be regarded with suspicion.
The rising of birds shows an ambush.
Startled beasts show that the enemy is stealthily approaching from several sides."

BEYOND THE IMPLICATIONS of treacherous terrains, Sun Tzu warns of the necessity of reading the environment for signs of the enemy's presence and intentions.

Confined spaces may conceal ambushes; a close but quiet opponent is potentially focused and ready to attack. Always be alert to unseen danger and attune yourself to subtle clues and threats disguised by surface calm. It's not just about reacting, but foreseeing as a result of clues hidden in plain sight.

“High, straight spurts of dust betoken that chariots are coming.

Long, low masses of dust show the coming of infantry.

Here and there, thin and high columns of dust are signs that firewood and fodder are being collected.

Small clouds of dust moving to and fro are signs that the enemy is preparing to encamp for a short time.

Busy preparations and smooth words show that the enemy is about to advance to attack.

Big words, and the spurring forward of horsemen, are signs that the enemy is about to retire.

An advance of the light chariots to the flanks of the camp is a sign that the enemy is coming forth to fight.”

HERE, SUN TZU warns of the need to be observant of subtle indications of the enemy's plans. He explains how the enemy's intentions can be observed through the details their presence reveals – the rising and formation of dust, for example, can alert us to unseen movements. Quite simply, a skilled leader reads the environment as a form of intelligence.

" Without consultation, suddenly to desire an armistice,
is a mark of ulterior design.
The passing to and fro of messengers, and the forming
up of troops, show that the enemy has some movement
on foot.
An advance, followed by sudden retirement,
is a lure to attack.
When the enemy use their weapons to rest upon,
they are hungry.
If the drawers of water drink at the river, the enemy is
suffering from thirst.
Disregard of booty that lies ready at hand is a sign of
exhaustion.
The clustering of birds round a position shows that it is
unoccupied.
Voices calling in the night betoken alarm.
Disorder in the army is a sign that the general is disregarded.
A changing about of flags and banners is a sign that the army
is unsettled.
If the officers be angry, it is because the soldiers are tired,
*and slow to obey.*
The killing of horses for food shows that the enemy is short
of provisions.
When the cooking-pots are hung up on the wall and the
soldiers turn not in again, the enemy is at an end of his
resources.
Exceeding graciousness and familiarity on the part of the
general show that he has lost the confidence of the soldiers.
Frequent rewards show that discipline is at an end.
Frequent punishments are a sign that the general
is in difficulties. "

In this passage, Sun Tzu continues the theme of military observation and psychological intelligence, advising that we look for external signs of our opponent's situation – their morale, supply situation, discipline and intentions.

When the enemy behaves in erratic or unexpected ways, they may be giving signs of their instability. A sudden request for peace, without prior discussion, is likely a trick – buying time, hiding weakness or setting up an ambush. Signs of hunger, thirst, anger or indiscipline can imply that morale is low and troops are unsettled and the leader is losing control. Here is your chance to identify the quiet symptoms of weakness in your opponent and press your advantage.

"The general who first blusters, and then is obsequious, is without perception. He who offers apologies and hostages is anxious for a truce. When both sides, eager for a fight, face each other for a considerable time, neither advancing nor retiring, the occasion requires the utmost vigilance and circumspection. Numbers are no certain mark of strength. Even if incapable of a headlong assault, if the forces be united, and the enemy's condition ascertained, victory is possible. He who without taking thought makes light of the enemy is certain to be captured."

Sun Tzu now turns his attention to the strategic importance of focus and consistency and signs of weak leadership. He suggests that a leader who is initially full of menace and protest but at a later point is fawning and ingratiating, is vulnerable and ready to being exploited. True leadership is consistent and calculated, not driven by emotion or appearance.

He further suggests that apologies and hostages are not peace offerings – they're signs of desperation. The enemy is weakened or afraid, seeking safety through diplomacy.

Sun Tzu goes on to alert us to the dangers of a prolonged standoff as this is when miscalculations and mistakes are easily made. Patience and alertness are crucial.

He cautions that strength is more than numbers. Good leadership, morale, coordination, knowledge of terrain and timing are of equal importance. Furthermore, if your troops are united, and you understand the enemy's strengths and weaknesses, "victory is still possible".

Victory belongs to those who stay united, think clearly and do not underestimate their opponent.

“ If a general who is strange to the troops punish them, they cease to obey him.
If they are not obedient, they cannot be usefully employed.
If the troops know the general, but are not affected by his punishments, they are useless.
By humane treatment we obtain obedience; authority brings uniformity.
Thus we obtain victory.
If the people have been trained in obedience from the beginning, they respect their leader's commands.
If the people be not early trained to obedience, they do not respect their leader's commands.
Orders are always obeyed, if general and soldiers are in sympathy. ”

In this passage, Sun Tzu focuses on the psychology of leadership and discipline, and explains how authority, familiarity, training and emotional connection all play a role in effective command.

He suggests that when a leader is unfamiliar to the people they command and leads with punishment, it creates fear and resentment, not respect. People will not follow someone they don't trust or recognize.

Conversely, if they do know the leader but that person is too lenient or fails to enforce consequences, discipline collapses. Fairness brings obedience, authority ensures order and coordinated action, and together they create a powerful, disciplined and cohesive force.

Sun Tzu advises that discipline is most effective when it's instilled right from the start and concludes that "Orders are always obeyed, if general and soldiers are in sympathy." In other words, emotional bonds and mutual respect ensure obedience.

## CHAPTER TEN

# GROUND

IN THE FIRST part of this chapter, Sun Tzu describes the six different types of terrain and advises on how and when to engage or not engage with the enemy on each type.

Having considered the *physical* terrain, Sun Tzu then turns his focus to what might be described as the *psychological* terrain, created by the leader. He describes "six calamities" that may arise as a result of faulty leadership, showing that a leader's weakness or indecision has the same impact on the soldiers' will to fight as the six kinds of terrain. These "six calamities" also apply to the officers in an army.

The chapter ends by explaining that the general has the right to overrule the sovereign's orders, and that, if he advances without coveting fame and retreats without fearing disgrace, and if he "only strives for the people's welfare, and his lord's advantage", he will become "a treasure to the state".

Furthermore, if he treats his soldiers well but firmly, understands and manages the soldiers and the terrain, knows the enemy and himself, victory is certain.

“With regard to the different natures of ground there are:—Open ground; broken ground; suspended ground; defiles; precipices; far countries.”

THIS SHORT PASSAGE introduces the six different natures of ground. They are:

1. Open ground: accessible land with plenty of roads and means of communications.

2. Broken ground: land that is difficult to access and traverse, where communications might break down and retreat may be difficult.

3. Suspended ground: land where neither side can gain by making the first move.

4. Defiles: narrow passes – gorges or tight corridors – where you're vulnerable to ambush.

5. Precipices: sheer cliffs or dangerous heights, where panic or missteps can be fatal.

6. Far countries: positions that are a great distance from the enemy.

" Open ground is that where either side has liberty of movement: be quick to occupy any high ground in the neighbourhood and consider well the line of supplies.

Broken ground. Advance is easy, but retreat from it is difficult. Here, if the enemy be not prepared, we may win: but should he be prepared, and defeat us, and retreat be impossible, then there is disaster.

Suspended ground. The side that takes the initiative is under a disadvantage. Here, if the enemy offer some allurement, we should not advance: but rather, by feigning retreat, wait until he has put forth half his force. Then we may attack him with advantage.

Defiles, make haste to occupy; garrison strongly and await the enemy. Should the enemy be before you, and in strength, do not engage him; but if there be unoccupied points, attack him.

In precipitous ground quickly occupy a position on a sunny height, and await the enemy. If the enemy be before you, withdraw and do not attack him.

If distant from the enemy, and the forces be equal, to take the initiative is disadvantageous.

Now, these are the six kinds of ground. It is the duty of generals to study them. "

Sun Tzu maps out wise ways of responding to the challenges presented by these terrains. He asserts that:

1. On open ground where both sides can move freely, the priority is to find high ground and secure supply lines to ensure one is never exposed or vulnerable.
2. On broken ground, caution is urged because a well-organized enemy may have a natural advantage.
3. Suspended ground represents great risk too – here taking action might be stepping into a trap, so patience may be the best option.
4. In defiles – narrow passes – it's important to occupy and reinforce the ground before your enemy does so.
5. On precipitous ground the enemy may already hold the advantage, so proceed only if you can secure the upper hand.
6. And in far countries, where action requires much effort and travel, it may be best to wait rather than go.

" Again, there are six calamities among
the troops, arising, not from defect
of ground, or lack of opportunity,
but from the general's incapacity.
These are: repulse, relaxation, distress,
disorganisation, confusion and rout.
If troops be sent to attack an enemy of
equal quality, but ten times their number,
they retire discomfited.
Strong soldiers with weak officers
cause relaxation.
Able officers with feeble soldiers
cause distress.
Enraged senior officers, who fall upon
the enemy without orders, and obey not
the general because he does not recognise
their abilities, produce disorganisation. "

HERE SUN TZU sheds light on a range of military failures that stem not from environment or terrain but from poor leadership; specifically, the leader's incompetence or mismanagement.

The "six calamities" are:

1. Repulse – being driven back or defeated.
2. Relaxation – lack of discipline and effort.
3. Distress – strain, breakdown under pressure.
4. Disorganization – lack of structure and coordination.
5. Confusion – unclear commands or chaotic execution.
6. Rout – a total collapse or panic-driven retreat.

The message is clear – the greatest dangers often come from within, not from the enemy. All six calamities reflect poor strategic judgment, a lack of discipline, mismatched leadership and capability and a fragmented leadership structure.

These ancient military principles have implications in our world today, in, for example, areas such as business, sports teams, politics or social movements. Failure in a number of situations is often not due to competitors, adversaries or other external forces but to poor leadership decisions and team dynamics.

" Weak and amiable generals,
whose directions and leadership
are vague, whose officers' and men's
duties are not fixed, and whose
dispositions are contradictory,
produce confusion.
Generals, who are unable to
estimate the enemy, who oppose
small numbers to large, weakness
to strength, and who do not put
picked men in the van of the army,
cause it to be routed.
These six things lead to defeat.
It is the duty of the general to
study them carefully. "

IN THIS PASSAGE, Sun Tzu outlines six key leadership failures that lead directly to confusion, inefficiency and defeat. Leadership is not just about being in charge – it's about creating clarity, strength and unity.

These timeless principles apply from battlefields to boardrooms.

" Ground is the handmaid of victory. Ability to estimate the enemy, and plan the victory; an eye for steepness, command and distances: these are the qualities of the good general. Whosoever knows these things, conquers; he who understands them not, is defeated. If victory be certain from the military standpoint, fight, even if the lord forbid. If defeat be certain from the military standpoint, do not fight, even though the lord commands it. The general who advances, from no thought of his own glory, or retires, regardless of punishment; but only strives for the people's welfare, and his lord's advantage, is a treasure to the state. "

SUN TZU REITERATES the characteristics of a good leader – an ability to understand the environment, to accurately assess the enemy and to plan wisely.

He reminds us that a true leader's actions are based on strategic reality, not strict obedience to the head of state, even causing them to defy orders if necessary. A leader who is selfless, putting the welfare of the people above personal glory or fear of punishment, is "a treasure to the state".

> The wisdom in this passage – that effective leadership depends on integrity, ethical decision making and responsibility – applies to our lives today in areas such as governance, business and sport.

“ The good general cares for his soldiers,
and lovingly treats them as his children;
as a consequence they follow him through
deep valleys, and are beside him in death.
Nevertheless, over-care for the soldiers may cause
disobedience; over-attention may make them
unserviceable; over-indulgence may produce
disorder: they become like spoilt children,
and cannot be used.
He who is confident of his own men, but is
ignorant that the enemy should not be attacked,
has no certainty of victory.
He who knows that the enemy may be attacked
with advantage, but knows not his own men, has no
certainty of victory.
Confidence in the troops, right judgment when
to attack the enemy, but ignorance of the ground,
bring uncertain victory.
The wise soldier, once in motion, does not waver,
and is never at a loss.
As has been said: ‘Know thyself; know the enemy;
fear not for victory.’
Also, if the season and the opportunity be
realised, and the ground known, complete
victory is certain. ”

IN THIS PASSAGE, Sun Tzu continues to explain the traits of a successful leader. A good leader treats soldiers with care and earns their loyalty, but not to the extent of indulging them and so causing disorder.

Victory requires balance – it depends not only on knowing one's troops but also on knowing the enemy, knowing when and where to attack, as well as understanding the terrain and timing. True success comes from self-knowledge, strategic awareness and decisive action – when these are aligned, victory is assured.

> These values remain true of today's leaders – whether in business, politics or sport. Effective leadership requires balance: caring for your team builds loyalty, but overindulgence can lead to lack of discipline and poor performance. When you know yourself and your opponent, understand the external environment, and act with clarity and timing, success is likely.

## CHAPTER ELEVEN

# THE NINE GROUNDS

THIS IS THE longest chapter in *The Art of War*, so here we focus on the "Nine Grounds" that give the chapter its title and which lie at its heart. The Nine Grounds are the nine different types of terrain on which a battle may take place, ranging from the most favorable to the most difficult.

In this chapter, Sun Tzu begins by briefly accounting for each of the nine types of ground and then expounds on how to handle each type.

" In respect to the conduct of war there are:—
Distracting ground; disturbing ground; ground of contention; intersecting ground; path-ridden ground; deeply-involved ground; difficult ground; enclosed ground; death ground.

[1] At all times, when the prince fights in his own territory, it is called distracting ground.*

[2] That ground a short way inside the enemy's border is called disturbing ground.

[3] Ground giving advantage to whichever side is in possession, is called ground of contention.

[4] Ground to which either side has access, is called intersecting ground.

[5] Ground between three provinces first possession of which enables the peoples of the earth to be controlled, is called path-ridden ground.

[6] The interior of the enemy's country with many of his fortified towns in rear, is called deeply-involved ground.

[7] Mountain and forest, precipices, ravines, marsh and swamp, all places where passage is hard, are called difficult ground.

[8] A narrow entrance and winding outlet, where a small number can oppose a large force, is called enclosed ground.

[9] That ground where delay means disaster, is called death ground. "

* This and the following are so called because the men are continually thinking of, and slipping back to, their homes.

Sun Tzu identifies nine types of terrain in warfare, each requiring different strategies:

1. Distracting ground – the army fights on its own territory, where emotional and political distractions may arise.
2. Disturbing ground – just inside enemy territory, causing unease and instability.
3. Ground of contention – valuable land that offers advantage to whoever holds it.
4. Intersecting ground – easily accessible by both sides, leading to frequent conflict.
5. Path-ridden ground – strategic crossroads affecting control over larger regions.
6. Deeply-involved ground – deep within enemy territory, surrounded by hostile forces.
7. Difficult ground – challenging terrain like mountains, forests and swamps.
8. Enclosed ground – tight spaces where small forces can resist larger ones.
9. Death ground – a critical situation where survival depends on immediate action – no retreat is possible.

Each type of ground demands a specific tactical response, emphasizing the importance of adapting strategy to environment.

"Wherefore, do not fight on distracting ground; do not linger on disturbing ground.
If the enemy be in possession of disputed ground, do not attack.
In intersecting ground, do not interrupt the highways.
At the crossing of highways, cultivate intercourse.
When deeply involved, levy and store up the enemy's property.
Quickly depart from difficult ground.
On enclosed ground, use stratagem.
On death ground, fight.
[...]"

IN THIS PASSAGE, Sun Tzu advises on how to manage conflict on the nine types of ground. Once again, he counsels that a leader should adapt their strategies to the nature of the terrain and specific conditions of the battlefield, while at the same time know their position and that of the enemy. By doing so, a leader can make informed decisions that increase the likelihood of success.

> Just as terrain affects military tactics, success in life comes not from rigid plans but from an awareness of your environment, flexibility and timely action.

" The different natures of the Nine Grounds; the suiting of the means to the occasion; the hearts of men: these are things that must be studied. When deep in the interior of a hostile country, there is cohesion; if only on the borders, there is distraction. To leave home and cross the borders is to be free from interference.

On distracting ground, unite the soldiers' minds.
On disturbing ground, keep together.
On disputed ground, try to take the enemy in rear.
On intersecting ground, look well to the defences.
On path-ridden ground, cultivate intercourse.
On deeply-involved ground, be careful of supplies.
On difficult ground, do not linger.
On enclosed ground, close the path of escape.
On death ground, show the soldiers that there is no chance of survival.

It is the nature of soldiers to defend when surrounded, to fight with energy when compelled thereto, to pursue the enemy if he retreat.

[...] "

HERE, SUN TZU shares insights in terms of both the nine types of terrain and human nature – "the hearts of men".

He explains that when troops are far into enemy territory they become more unified and focused – they must work together to survive. When they're near the border they feel less urgency and are more likely to be distracted or divided. However, once an army crosses into enemy land, it's out of reach of domestic politics and distractions, and can focus solely on its mission.

## CHAPTER TWELVE

# ASSAULT BY FIRE

IN THIS CHAPTER, Sun Tzu describes five ways of attacking by fire and explains the circumstances and preparations to be considered before making an attack.

Having described the use of fire as a physical weapon, he then turns to a series of statements that might be interpreted as describing the "fire" of a leader's approach and mindset. He points out, for example, that a leader must have a spirit of boldness and energy but at the same time must not act out of anger and frustration.

“There are five ways of attack
by fire:
The first is called barrack burning;
the second, commissariat burning;
the third, equipment burning; the
fourth, store burning; the fifth,
the company burning.”

Fire, in ancient warfare, was a powerful tool – not just for destruction but for causing chaos, lowering morale and disrupting supply lines. Each of the five methods targets a specific element of the enemy's military infrastructure to cause maximum disruption:

1. burning enemy soldiers in their camp
2. burning the enemy stores
3. burning their supply wagons
4. burning their weapons
5. hurling dropping fire among the enemy.

> "The moment for the fire assault must be suitable. Further, appliances must always be kept at hand. There is a time and day proper for the setting and carrying out of the fire assault; namely: such time as the weather is dry; and a day when the moon is in the quarters of the stars Chi, Pi, I, Chen:* for these are days of wind."

* Another Edwardian translator of *The Art of War*, Lionel Giles (1875–1958), translated these as the Chinese constellations The Sieve, The Wall, The Wing and The Cross-bar, "corresponding roughly to Sagittarius, Pegasus, Crater and Corvus".

SUN TZU EMPHASIZES that fire attacks should be meticulously planned, that one must have favorable means and materials available and ready for use. Consideration of weather conditions – particularly dry and windy days – is important in order to ensure the fire's effectiveness.

His comment "[...] when the moon is in the quarters of the stars Chi, Pi, I, Chen: for these are days of wind" refers to ancient Chinese astrology, where certain lunar positions (Chi, Pi, I, Chen are star constellations or lunar mansions) were believed to coincide with windy conditions. Wind was essential to spread fire effectively.

"Regard well the developments that will certainly arise from the fire, and act upon them. When fire breaks out inside the enemy's camp, thrust upon him with all speed from without; but if his soldiers be quiet, wait, and do not attack.
When the fire is at its height, attack or not, as opportunity may arise.
If the opportunity be favourable, set fire to the enemy's camp, and do not wait for it to break out from within.
When fire breaks out on the windward side, do not attack from the leeward.
Wind that rises in the day lasts long. Wind that rises in the night time quickly passes away.
The peculiarities of the five burnings must be known, and the calendar studied, and, if the attack is to be assisted, the fire must be unquenchable.
If water is to assist the attack, the flood must be overwhelming.
Water may isolate or divide the enemy; fire may consume his camp; but unless victory or possession be obtained, the enemy quickly recovers, and misfortunes arise. The war drags on, and money is spent."

IN THIS PASSAGE, Sun Tzu counsels for the need to be aware of the developments that will arise from a fire; that a leader must be fully conversant with the timing and the characteristics of the "five burnings".

There is further advice concerning the use of water to assist an attack and an assertion that with the use of fire and water, the effect must be decisive and final, otherwise "the enemy quickly recovers [...]. The war drags on, and money is spent".

By targeting the enemy's resources and morale through fire, a leader can weaken the opponent without direct confrontation, aligning with Sun Tzu's principle of winning without fighting.

" Let the enlightened lord consider well;
and the good general keep the main object in
view. If no advantage is to be gained thereby,
do not move; without prospect of victory,
do not use the soldiers; do not fight unless
the state be in danger.
War should not be undertaken because the lord
is in a moment of passion. The general must not
fight because there is anger in his heart.
Do not make war unless victory may be gained
thereby; if there be prospect of victory, move;
if there be no prospect, do not move.
For passion may change to gladness, anger
passes away; but a country, once overturned,
cannot be restored; the dead cannot be
brought to life.
Wherefore it is written, the enlightened lord is
circumspect, and the good general takes heed;
then is the state secure, and the army
victorious in battle. "

Sun Tzu emphasizes the importance of deliberate and rational decision making before embarking on warfare.

Military action should only be undertaken if the state is in danger. Otherwise it should be avoided. Decisions to go to war should not be driven by emotions such as anger or passion. Impulsive moves lead to unnecessary conflict and loss.

However, if you do go into battle, unless victory looks certain, it's wiser to avoid confrontation. The destruction caused by war – loss of life, resources and social upheaval – is permanent and cannot be reversed. A sensible ruler and a prudent leader have a responsibility to prioritize the state's security and the wellbeing of its people over personal emotions or desire for retribution.

## CHAPTER THIRTEEN

# THE EMPLOYMENT OF SPIES

SUN TZU BEGINS this chapter by reiterating a point made at the start of Chapter 2: a prolonged war is an expensive business; one way or another, nobody wins and everyone loses. What is needed is foreknowledge. Foreknowledge, says Sun Tzu, is "not to be got by calling on gods and demons; nor does it come of past experience nor calculation". Knowledge of the enemy's dispositions can only be obtained from other men – in other words, from secret agents and spies.

Sun Tzu advocates for the strategic use of spies to gain critical information about the enemy, thereby reducing the duration and cost of warfare. He tells us there are five types of spies: village (i.e. local) spies, inward spies, converted spies, death (i.e. doomed) spies and living (i.e. surviving) spies. Su Tzu explains how to manage each type of spy and concludes by explaining that knowledge of the enemy begins with the converted spy; that the converted spy not only brings information themselves but they make it possible to use the other kinds of spy to full advantage.

"Calling 100,000 men to arms, and transporting them a hundred leagues, is such an undertaking that in one day 1,000 taels of the citizens' and nobles' money are spent; commotions arise within and without the state; carriers fall down exhausted on the line of march of the army; and the occupations of 700,000* homes are upset.
Again, for years the armies may face each other; yet the issue may depend on a single day's victory.
Wherefore, by grudging slight expense in titles and salaries to spies, to remain in ignorance of the enemy's circumstances, is to be without humanity.
Such a person is no general; he is no assistance to his lord; he is no master of victory.
The enlightened ruler and the wise general who act, win, and are distinguished beyond the common, are informed beforehand.
This knowledge is not to be got by calling on gods and demons; nor does it come of past experience nor calculation. It is through men that knowledge of the enemy is gained."

* The population was divided, for military purposes, into groups of eight families. In time of war, each group sent one man into the field, furnished his wants and provided for his family. Therefore, if 100,000 men are taken, 700,000 homes are affected.

THIS PASSAGE EMPHASIZES the immense costs of warfare and the critical importance of intelligence in achieving victory. Sun Tzu reiterates the fact that war is a drain on the resources of the state; it's an expensive business. There will be political and social disturbance at home and abroad. Lives will be disrupted. Despite long campaigns, victory can hinge on a single day.

Therefore, neglecting to invest in spies and timely intelligence is irresponsible and inhumane. Wise leaders understand that success depends on accurate, human-sourced knowledge of the enemy – not on superstition or guesswork.

A leader who acts without proper knowledge isn't just ineffective – they're harmful to those who rely on them. Even in a digital age, people, conversations, observations, human insights and networks are vital.

“ Now the five kinds of spies are these: village spies, inner spies, converted spies, death spies, living spies.

If these five means be employed simultaneously, none can discover their working. This is called the Mysterious Thread: it is the Lord's Treasure.

[1] Village spies are such people of the country as give information.

[2] Inner spies are those of the enemy's officials employed by us.

[3] Converted spies are those of the enemy's spies in our pay.

[4] Death spies are sent to misinform the enemy, and to spread false reports through our spies already in the enemy's lines.

[5] Living spies return to report.

In connection with the armies, spies should be treated with the greatest kindness; and in dealing out reward, they should receive the most generous treatment. All matters relating to spies are secret. ”

SUN TZU OUTLINES five distinct types of spies. When employed effectively and in unison, these spies form a covert network referred to as the "Mysterious Thread", an invaluable asset for any ruler or leader.

1. Village spies (local spies): their familiarity with the area makes them valuable sources of intelligence.
2. Inner spies: these officials within the enemy's administration can offer insights into enemy plans, policies and internal dynamics.
3. Converted spies (double agents): these can be used to feed false information back to their original side or to identify other enemy operatives.
4. Death spies: these spies are doomed; they are put to death when the enemy finds out that he has been tricked.
5. Living spies: these are operatives who gather intelligence and return safely to report their findings.

Recognizing the vital contributions of spies, Sun Tzu advocates for their generous treatment. He says that they should be well rewarded and treated with utmost respect to ensure their loyalty and effectiveness. Matters concerning spies must be handled with the highest level of secrecy to protect their identities and the integrity of the intelligence network.

" Without infinite capacity in the general, the employment of spies is impossible.
Their treatment requires benevolence and uprightness.
Except they be observed with the closest attention, the truth will not be obtained from them.
Wonderful indeed is the power of spies.
There is no occasion when they cannot be used.
If a secret matter be spoken of before the time is ripe, the spy who told the matter, and the man who repeated the same, should be put to death. "

THE USE OF spies must be well managed. A leader must have the ability to interpret intelligence accurately and the discernment to make strategic decisions based on that information.

Spies must be managed with goodwill, sincerity and honesty. However, a leader is well advised to keep their wits about them; to be aware that a spy's reports may not be reliable or true.

Sun Tzu tells us that the use of spies is not limited to specific situations; their utility spans across all phases of military operations, providing critical insights that can influence the outcome of conflicts.

Finally, if a spy discloses information to another, they must both be put to death to prevent the information leaking any further.

"If desirous of attacking an army; of besieging a fortress; or of killing a certain person; first of all, learn the names of the general in charge; of his right-hand men; of those who introduce visitors to the Presence; of the gate keeper and the sentries. Then set the spies to watch them. Seek out the enemy's spies who come to spy on us; give them money; cause them to be lodged and cared for; and convert them to the service. Through them we are enabled to obtain spies among the enemy's villagers and officials.

By means of the converted spy, we can construct a false story for the death spy to carry to the enemy."

Two matters are addressed by Sun Tzu in this passage. Firstly, targeted intelligence gathering. This involves identifying the names and roles of the enemy's leader, their close aides, gatekeepers and sentries. Your spies can then be used to monitor these individuals, gathering information on their routines, strengths and vulnerabilities. This approach ensures that decisions are based on accurate, detailed intelligence, maximizing your chances of a successful operation.

The second issue is the matter of converting enemy spies. Once they have been identified, they should be offered bribes, comfortable housing and respectful treatment – to entice them to defect. Not only can converted spies provide insights into the enemy's operations, they will also know which local people are open to gain and which officials open to corruption. The converted spy knows how the enemy can best be deceived and so will feed the death spy false information that will be carried back to the enemy.

Converting enemy spies not only neutralizes a threat but also transforms it into a valuable asset.

“It is through the converted spy that we are able to use the five varieties, to their utmost advantage; therefore he must be liberally treated.

[…]

Wherefore, intelligent rulers and wise generals use the cleverest men as spies, and invariably acquire great merit. The spy is a necessity to the army. Upon him the movement of the army depends.”

In this final passage, Sun Tzu asserts that the converted spy is pivotal in creating an effective intelligence network. They not only bring information themselves, they make it possible to use the other kinds of spy to advantage. Hence it's essential that the converted spy be treated with the utmost goodwill and generosity.

> Wise leaders rely on the most capable individuals as spies and treat them well, recognizing that the army's movements and outcomes often depend on the information they provide.
>
> The ancient commentator Chia Lin sums up all this in a striking image: "An army without spies is like a man without ears or eyes."

Also available as part of the Timeless Wisdom series

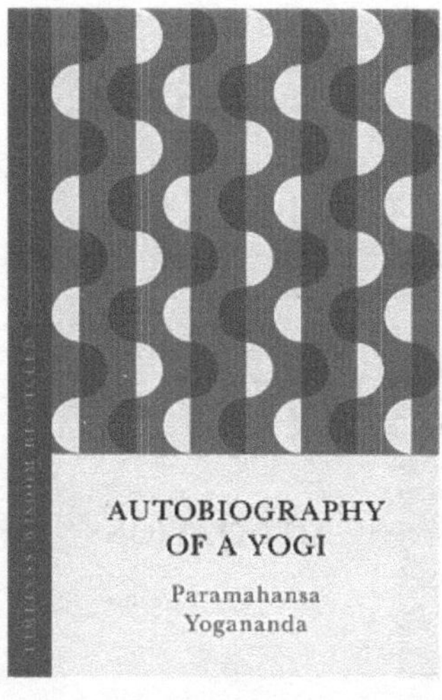


***Autobiography of a Yogi: Timeless Wisdom Distilled* by Paramahansa Yogananda**

*Embark on a journey of spiritual discovery and inner transformation.*

In his extraordinary *Autobiography of a Yogi*, Paramahansa Yogananda shares a visionary account of his life, teachings and encounters with saints, sages and seekers. A gateway into India's spiritual heritage, it is one of the most influential memoirs of the 20th century.

This distilled edition features the most powerful teachings from the original text, accompanied by illuminating commentary from bestselling author Gill Hasson. Deeply inspiring and accessible, it invites you to explore a life of purpose, connection and self-realization.

Hardback ISBN: 978 1 399 82152 0
Ebook ISBN: 978 1 399 82153 7

For more information, please visit www.johnmurraypress.co.uk

Also available as part of the Timeless Wisdom series

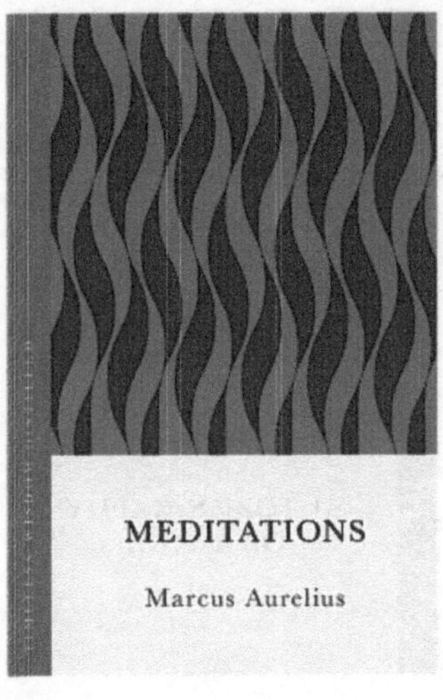


*Meditations: Timeless Wisdom Distilled* by Marcus Aurelius

*Find clarity, resilience and purpose in turbulent times.*

Roman emperor and Stoic philosopher Marcus Aurelius's *Meditations* is a profound and personal guide to living with integrity, courage and calm. His reflections remain strikingly relevant, speaking across the centuries to anyone seeking a more meaningful and grounded life.

This distilled edition presents his most powerful insights, accompanied by fresh, accessible commentary from bestselling author Gill Hasson. A wise and practical companion for everyday life, this book will help you build inner strength and live in accordance with your highest values.

Hardback ISBN: 978 1 399 82148 3
Ebook ISBN: 978 1 399 82149 0

For more information, please visit www.johnmurraypress.co.uk

Also available as part of the Timeless Wisdom series

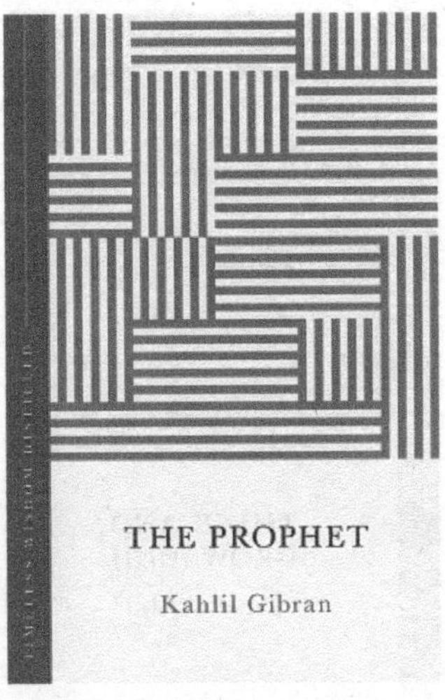


*The Prophet: Timeless Wisdom Distilled* by Kahlil Gibran

*Let poetry open the door to wisdom, beauty and inner peace.*

Kahlil Gibran's *The Prophet* is a luminous meditation on love, freedom, sorrow and joy. These poetic essays capture life's deepest truths with grace and simplicity, sharing timeless insight into the human spirit.

This distilled edition brings together the most enduring passages, supported by thoughtful reflections from bestselling author Gill Hasson. A soul-stirring companion for reflection, it's a book to return to again and again for guidance and inspiration.

Hardback ISBN: 978 1 399 82154 4
Ebook ISBN: 978 1 399 82155 1

For more information, please visit www.johnmurraypress.co.uk

Also available as part of the Timeless Wisdom series

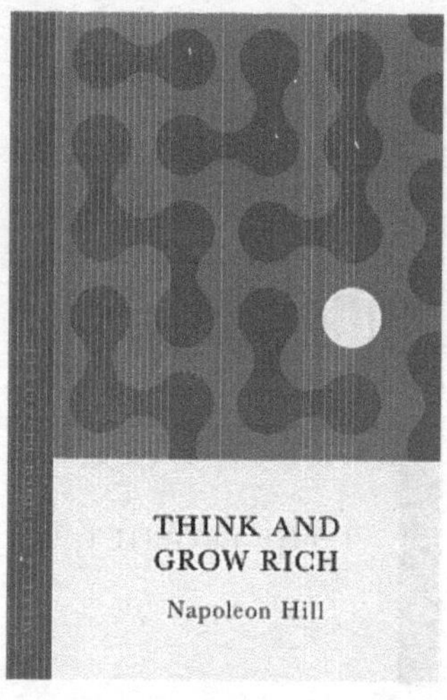


*Think and Grow Rich: Timeless Wisdom Distilled* by Napoleon Hill

*Discover the power of your thoughts to shape your destiny.*

*Think and Grow Rich* is Napoleon Hill's timeless guide to achieving personal and financial success. Drawing on the insights of history's most successful figures, it sets out a practical blueprint for turning desire into achievement through belief, persistence and focused thought.

This distilled edition features all of the original's most powerful lessons, paired with fresh and revealing commentary from bestselling author Gill Hasson. Clear, compelling and deeply relevant today, it's your essential guide to unlocking your potential and creating the life you want.

Hardback ISBN: 978 1 473 63626 2
Ebook ISBN: 978 1 473 63627 9

For more information, please visit www.johnmurraypress.co.uk

**RAISING READERS**

**Books Build Bright Futures**

Dear Reader,

We'd love your attention for one more page to tell you about the crisis in children's reading, and what we can all do.

Studies have shown that reading for fun is the **single biggest predictor of a child's future life chances** – more than family circumstance, parents' educational background or income. It improves academic results, mental health, wealth, communication skills, ambition and happiness.[1]

The number of children reading for fun is in rapid decline. Young people have a lot of competition for their time. In 2024, 1 in 10 children and young people in the UK aged 5 to 18 did not own a single book at home.[2]

Hachette works extensively with schools, libraries and literacy charities, but here are some ways we can all raise more readers:

- Reading to children for just 10 minutes a day makes a difference
- Don't give up if children aren't regular readers – there will be books for them!
- Visit bookshops and libraries to get recommendations
- Encourage them to listen to audiobooks
- Support school libraries
- Give books as gifts

There's a lot more information about how to encourage children to read on our website: **www.RaisingReaders.co.uk**

Thank you for reading.

hachette UK

---

[1] National Literacy Trust, Book Ownership in 2024, November 2024 https://nlt.cdn.ngo/media/documents/Book_ownership_in_2024

[2] OECD. 2021. 21st-century readers: developing literacy skills in a digital world. Paris, France: OECD Publishing. https://www.oecd.org/en/publications/21st-century-readers_a83d84cb-en.html